TABLE OF CONTENTS

1. Introduction

2. Preparation, Initial Characterization, and Bottling

3. Bottle-to-Bottle Homogeneity Testing

4. Discussion of the Certification Method Developed for SRM 2881

5. Mass Axis Calibration Procedure

6. Signal Axis Calibration Procedure

7. Experimental Method for the Calibration of the Signal Axis

8. Data Analysis Methods

9. Experimental Results

10. Discussion of Type A Uncertainties

11. Discussion of Type B Uncertainties

12. Discussion of Certified Values

13. Conclusions

Acknowledgements

References

Figures

Tables

Appendix A SRM Certificate

1. Introduction

The certification of the absolute molecular mass distribution (MMD) of an *n*-octyl-initiated, proton-terminated, low mass, atactic polystyrene (PS) by matrix-assisted laser desorption ionization time-of-flight mass spectrometry (MALDI-TOF MS) is fully described. This polymer has been designated Standard Reference Material (SRM) 2881. The material constituting SRM 2881 was prepared commercially for the NIST Polymers Division specifically for use as a narrow-distribution polymer standard. Anticipated uses include calibration of mass spectrometers and size exclusion chromatographs (SEC) in accordance with various international standards (described below), as well as any other application where quantitative knowledge of the entire MMD is necessary. The work described here, and in related references, provides a means to create absolute molecular mass distribution standards of any low mass, narrow polydispersity polymer including proprietary materials.

This work is the culmination of a multi-year effort that required development of new methods in sample preparation[1], numerical instrument optimization[2], and unbiased data analysis[3-5]. This report mainly covers the final uncertainty determination and discusses a Taylor's expansion methodology to certify the absolute MMD of polydisperse samples.

2. Preparation, Initial Characterization, and Bottling

2.1 Synthesis

The polystyrene used for this SRM was prepared commercially by Scientific Polymer Products, Inc. (Ontario, NY) by anionic polymerization of styrene initiated with *n*-octyl lithium (FMC Lithium, Gastonia, NC) and terminated with a proton. From the preparation chemistry, we expected the polymer to be atactic polystyrene of the form:

$$CH_3-(CH_2)_7-[CH_2-CHPh]_n-CH_2-CH_2-Ph \qquad [2.1]$$

where Ph refers to the pendant phenyl group. The material was received in the form of a clean, white powder.

2.2 Initial Homogeneity and End-Group Composition Testing

Before certification, the within-lot homogeneity and approximate MMD of the candidate polymer were verified by size exclusion chromatography (SEC) to be what was requested by NIST and claimed by the manufacturer. Its end-group composition and number-average molecular mass were obtained by nuclear magnetic resonance (NMR). NMR found the expected *n*-octyl and proton end groups, an M_n in agreement with the value found by SEC, and indications of a trace (< 1 % by mass) of an aromatic solvent, likely a residual of the synthesis. Since the certification of SRM 2881 depends on comparisons between gravimetric and mass spectrometric compositions of polymer mixtures, all polystyrenes used were vacuum dried at 60 °C for 72 h to remove excess solvent.

From the above results we concluded the polymer was of the correct molecular structure and of the necessary purity to serve as a candidate for certification.

2.3 Bottling and Sampling

A total of 69 individual samples of SRM 2881 were prepared in screw-cap glass vials each containing about 0.25 g of polymer. The vials were divided into five subsets. One vial was randomly selected from each subset for bottle-to-bottle homogeneity testing of the packaged material. The first and last bottles were also taken for study. Furthermore, one bottle (containing about 2.5 g) was retained for use in future experiments. This bottle was also sampled for homogeneity as a matter of course.

3. Bottle-to-Bottle Homogeneity Testing

Bottle-to-bottle homogeneity testing was accomplished using size-exclusion chromatography (SEC). An Alliance 2000 GPC Liquid Chromatograph (Waters Corp., Milford, MA) with a differential refractive index detector and two Styragel 300 mm x 7.5 mm ID 10 μm HT6-E columns, and one Styragel 300 mm x 7.5 mm ID 10 μm HT-2 column (Waters Corp., Milford, MA) was used in this study. The chromatography was run at a 1.0 mL/min solvent flow rate. The injector and column compartment of the chromatograph were controlled at 40 °C for all measurements. Tetrahydrofuran (Mallinckrodt Specialty Chemicals, Paris, KY) with added antioxidant, 2,6-di-tert-butyl-4-methyl phenol (commonly known as butylated hydroxytoluene or

BHT), was used as the solvent. The addition of 5 μL of toluene per mL of THF created an SEC pump marker.

Two polymer solutions in tetrahydrofuran were made from each polymer sample vial. The polystyrene samples were dissolved in the solvent at a concentration of approximately 1.5 mg/mL. The order of preparing the solutions and the order of running the chromatograms was randomized [6]. SEC was performed using two injections from each solution.

After baseline subtraction, the SEC chromatograms were normalized to unit peak height and compared by simple overlaying to decide initially if there were visible differences outside the noise. The chromatograms from different solutions all superimpose on each other. This preliminary comparison showed that polymer samples taken from all the vials produced qualitatively identical chromatograms. In the next section, statistical analysis on the chromatography confirms quantitatively these initial visual observations.

3.1 Statistical Method to Compare Chromatograms: the Match Factor

In previous SRM homogeneity testing employing SEC a numerical match factor was used to compare one chromatogram with all the others. In this study, the match factor for chromatogram i is defined as the correlation coefficient between chromatogram i and the average chromatogram of the entire testing series. The match factor defined by Huber [7] as

$$MatchFactor = 10^3 \left\{ \Sigma xy - \frac{(\Sigma x \cdot \Sigma y)}{p} \right\}^2 \bigg/ \left[\left\{ \Sigma x^2 - \frac{(\Sigma x \cdot \Sigma x)}{p} \right\} \left\{ \Sigma y^2 - \frac{(\Sigma y \cdot \Sigma y)}{p} \right\} \right]$$

[3.1]

is simply the cross-correlation of two chromatograms. The value of x is the measured signal in the ith chromatogram and y is the measured signal from the average of all chromatograms at the same elution time; p is the number of data points in the chromatogram. The sums are taken over all data points in a region near the polymer peak of the chromatogram. At the extremes, a match factor of zero indicates no match while 1000.0 indicates an identical chromatogram. Generally, values above 999.0 indicate that the chromatograms are almost identical. Values between 900.0

and 999.0 indicate some similarity between chromatograms, but the result should be interpreted with care. All values below 900.0 are interpreted as an indication of different chromatograms [7].

For each polymer the chromatograms were run in groups of eight solutions on different days. Match factors against the average chromatogram of all the runs were obtained for the region of the analyte main peak. All chromatograms from the analyte main peak in this study had match factors against the mean chromatogram of greater than 999.0. An ANOVA (Analysis of Variance) calculated using the NIST interactive software for statistical and numerical data analysis called PC-OMNITAB (see NISTIR 4957) made on the match factors obtained from the chromatograms indicated that the match factors of chromatograms from vials of SRM 2881 were not different using a significance level of 0.05.

From the above considerations, we conclude that the vials of SRM 2881 are indistinguishable from one another.

4. Discussion of the Certification Method Developed for SRM 2881

Synthetic polymers are never obtained as a single molecular mass but rather as a distribution of molecular masses. The molecular mass distribution (MMD) and the mass moments (MM) derived from it are used in areas of polymer science as diverse as fundamental physical studies, polymer processing, and consumer product design. For example, the observed MMD is often compared to predictions from kinetic or mechanistic models of the polymerization reaction itself [8, 9]. Similarly, the tensile strength, melt flow rate, and many other physical properties are dependent on MMD. Determination of the MMD is used for quality control in polymer synthesis and as specification in international commerce. MALDI-TOF mass spectrometry to obtain the MMD of synthetic polymers has been integrated into various international standards such as ASTM D7034-05 entitled "Standard Test Method for Molecular Mass Averages and Molecular Mass Distribution of Atactic Polystyrene by Matrix Assisted Laser Desorption/Ionization (MALDI) Time of Flight (TOF) Mass Spectrometry (MS)". Similar standards are available from Deutsches Institut für Normung (DIN 55674) and the International Organization for Standardization (ISO WD 10927). SRM 2881 is anticipated to be a useful material for laboratories to check their proficiency in applying these standard methods.

Various averages, or molecular moments, can be defined as useful summaries of the MMD. Measuring and computing these summary statistics has comprised a major part of traditional polymer analysis. The three most common measures of the MMD are the number average relative molecular mass, M_n, the mass average relative molecular mass, M_w and the z-average relative molecular mass, M_z.

$$M_n = \sum_j m_j n_j / \sum_j n_j \qquad [4.1]$$

$$M_w = \sum_j m_j^2 n_j / \sum_j m_j n_j \qquad [4.2]$$

$$PD = M_w / M_n \qquad [4.3]$$

$$M_z = \sum_j m_j^3 n_j / \sum_j m_j^2 n_j \qquad [4.4]$$

where m_j is the mass of a discrete oligomer, n_j is the number of molecules at the given mass m_j. The experimental moments from MALDI-TOF MS are defined as M_n, M_w, and M_z, while the "true" values are given as M^0_n, M^0_w, and M^0_z. PD defines the polydispersity index which is a measure of the breadth of the polymer distribution. When PD is equal to one (i.e., in statistical terms the variance of MMD is zero), all of the polymer molecules in a sample are of the same molecular mass and the polymer is referred to as monodisperse. Natural polymers, such as proteins, are typically monodisperse. This is seldom the case for synthetic polymers, which have a non-point-mass MMD, whose width depends in a complex fashion on the polymerization chemistry, specifically its mechanism, kinetics, and reaction conditions. In general, no two polymerization reactions yield precisely the sample MMD no matter how carefully conditions are controlled. This means that the product each polymerization reaction must be measured to obtain a reliable MMD.

From classical polymer methods of analysis of the MMD one obtains one or two moments of the MMD. M_w can be obtained by light scattering or ultracentrafugation, and M_n can be obtained by the measurement of colligative properties like osmotic pressure, or by end group analysis as by nuclear magnetic resonance (NMR) or titration techniques. Generally these moments give an incomplete description of the overall MMD. Properties like melt viscosity, tensile strength, and impact strength often depend on the tails of the MMD rather than the central portion as defined by the two central moments. The polydispersity index can be used as a poor surrogate in determining the effect of the tails of the MMD. Thus, it is critically important that we make an effort to obtain the entire MMD. Furthermore, it is not uncommon, due to purposeful blending or to the polymer chemistry (resulting from two different mechanisms (intentionally of unintentionally) competing during preparation), that the polymer MMD will be bimodal; thus the central moments, M_n and M_w, are exceedingly poor representations of such MMD. Size exclusion chromatography (SEC) and matrix-assisted laser desorption/ionization time-of-flight mass spectrometry (MALDI-TOF MS) each offers the prospect of obtaining the entire MMD. SEC is not an absolute technique in that it always requires calibration with another polydisperse polymer analyte. Also, it is difficult to determine the Type B (systematic) uncertainty of an SEC measurement. MALDI-TOF MS, particularly at molecular masses below about 25 000 u for polystyrene (PS), can obtain a molecular mass spectrum (MMS) with distinct oligomer peaks. Such equivalent averages as described above are then applied to MMS of polymers where it is assumed that the area under a peak of an oligomer is proportional to the number of molecules at a given mass m_j. From the MMS one can, with proper understanding of the limitations and corrections of the spectrum, derive a good approximation to the true MMD of the polymer. It is the central purpose of this work to consider how one can use the MMS to approximate the MMD. For this we shall need to consider carefully the repeatability, accuracy, and precision of MALDI-TOF MS for polymer analysis.

One can ask, what is the precision and accuracy of the method? How well does the method measure the quantities it is purported to measure (in this case the MMD)? From one point of view, this can be rephrased as, how well does it agree with other methods measuring the same quantity? These other values can be obtained from an independent measurement on the sample using a separate method, or by comparing against a certified value measured using the same method. We call these methods "soft quantitation".

From a second point of view we can ask, how well does the method measure the "true" value of the quantity we are trying to measure? How we know what is the true value is often unclear, so we ask another question. What are the uncertainties in the measurement, other than statistical, which affect the measurement or the quantity we are trying to measure? We would like to be able to say something quantitative about these Type B (systematic) uncertainties in the MMS measured by MALDI-TOF MS arising from instrument attributes such as laser energy or detector voltage, and sample preparation parameters (such as choice of MALDI matrix, or matrix:salt:polymer ratio). In order to estimate these uncertainties, one needs a physical or mathematical model describing the experiment, i.e., how the experiment relates to the measurand. This is called "hard quantitation".

In this work the ratios of gravimetric masses of distinct polymers, defined as G_{gi}, for polymer i are compared to G_{Mi}, the total mass estimated from the MALDI-TOF MS molecular mass spectrum of polymer i. We cannot obtain the exact measure of total mass from MALDI-TOF MS but we can accurately and precisely measure mass ratios, i.e. the ratio G_{M1}/G_{M2}. Comparison of G_{g1}/G_{g2} to G_{M1}/G_{M2} is the rationale on which SRM 2881 is based.

5. Mass Axis Calibration Procedure

Mass axis calibration is more easily performed than signal axis calibration. Calibration of most time-of-flight instruments is usually done with monodisperse biopolymers of known molecular masses. These biopolymers are selected because they typically provide a single major peak whose mass is known accurately; thus, mass axis quantification is quite straightforward. Calibration usually can be done using three of these biopolymers. Collecting data with 2 ns time intervals, one can get better than single mass unit accuracy on an instrument with a 1.5 m flight tube in reflectron mode at a mass of about 7000 u.

In this work calibration of the mass axis was done by combining a single biopolymer with a homopolymer calibrant. The oligomeric masses, m_j, with n repeat units of mass r and masses of the end group, m_{end}, of the polymer calibrant are given by:

$$m_j = nr + m_{end} + m_{salt} \qquad [5.1]$$

where n is the number of repeat units in the n-mer of the polymer and the m_{salt} refers to the mass of the metal cation adducted to the polymer n-mer.

Calibration of the mass axis was accomplished through use of the biopolymer mass as follows. The main peak from the biopolymer bovine insulin is assigned to its mass of 5 730.61 u. The biopolymer peak will either lie between the masses of two n-mers of the polymer calibrant, or exactly correspond to the mass of an n-mer. If it is at exactly the same mass as one of the n-mers of the polymer calibrant, use equation [5.1] to find the degree of polymerization, n, for the n-mer. If the peak of the biopolymer lies between the masses of two n-mers of the polymer calibrant, use equation [5.1] to find n_1, the mass of the n-mer closest to the mass of biopolymer. Next we find additional calibration points by selecting polymer peaks at intervals between five and 10 repeat units both less than and greater than n_1 and compute their masses from equation [5.1]. Generally, a total of four or five calibration masses were selected.

For this work, mass accuracy of only a few mass units was necessary because the mixtures of polydisperse homopolymers we used had different end groups with masses differences on the order of tens of mass units (discussed below). All that was required is calibration close enough to positively identify each oligomer. Furthermore, the main uncertainty in converting from the MMS to the MMD lies in the signal axis quantitation. Achieving greater mass axis accuracy nets no increase in the accuracy of the MMD measurement.

6.0. Signal Axis Calibration Procedure

Most of the systematic uncertainties in converting from the MMS to the MMD arise in the signal axis calibration. Uncertainties arise from inconsistent, irreproducible sample preparation (particularly when using hand spotting methods), from drift in the machine parameters (especially variability in the laser energy over various time scales), from unequal detector sensitivity as a function of ion mass, from *ad hoc* data analysis methods (primarily background subtraction and peak integration), and from many other minor effects. With all these complications it seems impossible that we can find a simple way to look at quantitation on the signal axis; however, by taking the clue from many published papers (discussed below) that the polydispersity of a polymer must be fairly low to get a proper MMD from the MMS, a mathematical model was created to build on this observation. This entailed gravimetric blending of narrow polydispersity polymers to control the mixture polydispersity and discover a limiting case for narrow polydispersity materials. The literature in this area will be reviewed in the next section.

6.1 Gravimetric Blending for Signal Axis Quantitation

There are several works in the literature that use blending of polymers of different molecular mass to obtain some estimate of how polydispersity affects the predicted versus found properties of the MMD or moments thereof. Shimada *et al.*[10, 11] fractionated low molecular mass polystyrenes and polyethylene glycols into a single oligomers using preparative supercritical fluid chromatography (SFC). Oligomers with degree of polymerization up to 25 were fractionated. Then with five or six isolated oligomers from a given polymer they obtain a calibration curve to convert the MMS to the MMD as a function of machine parameter. This is certainly unique but depends on being able to fractionate the polymers into oligomers at any molecular mass, which as of this writing is only true for low mass oligomers. Furthermore, as we shall note later, it is important to be working under experimental conditions where there is a linear relationship between signal intensity and analyte concentration. As will be shown, this is necessary to make quantitative measurements of the MMD. Linearity was not demonstrated in the work of Shimada, *et al*.

Yan et al.[12] have looked at mass ratios of two polydimethylsiloxane polymers (average masses 2200 u and 6140 u) by MALDI-TOF MS. After initially optimizing the laser power to give the strongest signal intensity for both the low and high molecular mass polymers, they found that the mass ratios estimated from MALDI-TOF MS were very close to those predicted from gravimetric ratios. Chen, et al.[13,14] performed a similar experiment using polyethylene glycols with a variety of end groups. They found for molecular masses from 800 u to 3300 u the gravimetric amount of polymer seemed to map well into the MALDI-TOF MS measured values.

Li and coworkers [15-17] started by looking at equimolar blends of polystyrenes with molecular masses below 20 000 u in order to optimize the instrument and sample preparation parameters. They did this by varying machine parameters as well as matrix and salt ratios and finding where the M_n best matched the M_n expected of the equimolar mixtures. They made a careful study of PS of molecular mass 5050 u, 7000 u and 11 600 u and blends of the three. Fitting their polymer MMD to Gaussian distributions and using 5050 u and 11 700 u samples, they studied changes in the MMD of the polymer of 7000 u with blends of the 5050 u polymer. They found they could detect no systematic uncertainties within 0.5 % in the MMD or the moments of the MMD assuming MMD were Gaussian distributions.

6.2 Mathematical Model for Gravimetric Blending

As discussed in Section 4, to estimate the Type B uncertainty in an instrumental method one needs a mathematical model based on either physical principals or comprehensive phenomenological observations. A quantitative, predictive physical model of the MALDI process and the TOF mass spectrometer seems out of reach at this time. However, a heuristic mathematical model is available and will serve as our starting point. The experimental works described previously [12-17] each implicitly assume that there is a point in the parameter space of the instrument and the sample preparation where the signal intensity, S_i, for an oligomer of mass m_i is linearly proportional to n_i, the number of polymer molecules at that oligomer mass. Mathematically this is given by:

$$S_i = k n_i \qquad [6.1]$$

where for a narrow enough range of m_i we assume k is a constant independent of m_i and the range of linearity, $0 < n_i < n_0$, is the same for all oligomers. From the work of Goldschmidt and Guttman,[18] working with a 7900 u PS in a matrix of either retinoic acid or dithranol, there was found a linear relationship between signal intensity and analyte concentration over a wide concentration range provided there is a large molar excess of Ag salt. Furthermore, at a high polymer-to-matrix ratio, the curve becomes nonlinear and the signal intensity approaches saturation. This cutoff was shown to be a weak function of molecular mass.

If we assume we are in the linear region for all the oligomers of a polymer, the total measured signal is given by:

$$\sum S_i m_i = k \sum n_i m_i \qquad [6.2]$$

with $n_i m_i$ summed over all oligomers i. This is valid as long as the polymer is evenly distributed throughout the matrix on the target spot. Then we can obtain:

$$\sum S_i m_i / \sum S_i = k \sum n_i m_i / k \sum n_i \qquad [6.3]$$

The right hand side of the equation is the M_n of the polymer (equation [4.1]) independent of k since the k in numerator and denominator cancel out. The same holds for equations for M_w and all higher moments and is broadly true if we are in the linear range.

We know from the work of Montaudo and coworkers [19], McEwen and coworkers [20], and Li and coworkers [15-17] that if the m_i span too great a mass range the values for k and/or the n_0 cutoff must change dramatically otherwise the MALDI-TOF MS would be able to obtain the MMD correctly for very broad distributions.

In general, if we are in the linear range for each oligomer i of a polymer then:

$$S_i = k_i n_i \qquad [6.4]$$

where k_i is now a slowly varying function of i or more simply a slow varying function of oligomer mass m_i for a fixed set of instrument parameters and sample preparations parameters (e.g., matrix material, matrix:salt:polymer concentration, method of sample preparation, etc.). An equation similar to [6.4] has been assumed by Sato, *et al.* [21] for the relationship between fractions obtained from an analytical SEC and signals measured from each fraction by MALDI-TOF MS. Theirs is simply a coarse grained version of our mathematical model.

Now if we assume in the above equation that k_i is a slowly varying function of i or of m_i then we may make a Taylor's expansion around a mass peak near the center of the MMD, termed M_0. The center of the MMS is used to assure that the function is changing as little as possible over the entire width of the MMD. Then:

$$S_i = k_o n_i + Q(m_i - M_o)n_i + \textit{higher order terms in } n_i \textit{ and } m_i \qquad [6.5]$$

Here Q is a function of all the experimental conditions: the instrument parameters, the sample concentrations, and the sample preparation method. In the experimental procedure described later, once the instrument parameters and experimental preparation methods are optimized, every attempt is made to keep them constant to insure experimental reproducibility. Later is will be shown how variation in the machine parameters can affect the variation of Q and thus the Type B uncertainty. It should be noted that Q is a function of M_0, and is sometimes designated as $Q(M_0)$.

We shall now explore equation [6.5] and see how small linear shifts of the calibration constant Q over limited mass ranges effects quantities derivable from MALDI-TOF MS data. First we shall explore quantities like the total signal, the total detected mass, and the mass ratios of mixtures and see how these quantities relate to the true MMD of the polymer.

The total signal (S_T) from the polymer is given by:

$$S_T = \sum_i S_i = k_o \sum_i n_i + Q(M_n^0 - M_o) \sum_i n_i \qquad [6.6]$$

while the total mass of polymer detected (G_T) is given by:

$$G_T = \sum_i m_i S_i = k_o M_n^0 \sum_i n_i + Q M_n^0 (M_n^0 - M_o) \sum_i n_i \qquad [6.7]$$

where M^0_n and M^0_w are defined in equations 4.1 and 4.2, and are the true number average and mass average relative molecular masses. By multiplying equations 4.1 and 4.2 together we have:

$$M_w M_n = \sum_i m_i^2 n_i / \sum_i n_i \qquad [6.8]$$

Then taking the ratio of equations [6.6] and [6.7], we obtain:

$$M_n^{exp} = \sum m_i S_i / \sum S_i \qquad [6.9]$$

with the result that:

$$M_n^{exp} = M_n^o \left\{ \frac{(1+(Q/k_o)(M_w^o - M_o)}{(1+(Q/k_o)(M_n^o - M_o)} \right\} \qquad [6.10]$$

where M_n^{exp} is the experimentally-measured M_n.

For use later in this paper we obtain by the same algebra:

$$M_w^{exp} = \sum m_i^2 S_i / \sum m_i S_i \qquad [6.11]$$

with the result that:

$$M_w^{exp} = M_w^o \left\{ \frac{(1+(Q/k_o)(M_z^o - M_o))}{(1+(Q/k_o)(M_w^o - M_o))} \right\} \qquad [6.12]$$

All higher moments may be obtained in a similar way and have a similar form.

This then gives by simple division:

$$M_w^0 = M_w^{exp} \left\{ \frac{(1+(Q/k_o)(M_w^o - M_o))}{(1+(Q/k_o)(M_z^o - M_o))} \right\} \qquad [6.13]$$

which yields:

$$M_w^0 = M_w^{exp} \left\{ 1 - \frac{(Q/k_o)M_w^o(PD_w - 1)}{1+(Q/k_o)(M_z^o - M_o)} \right\} \qquad [6.14]$$

Let us consider this final equation. It says the deviation of the mass moment measured by MALDI-TOF MS from the true mass moment is a function of the polydispersity (PD) (arising from that moment) divided by a correction term arising from how far that moment is from the mass M_0 around which the Taylor's expansion to obtain k_0 and Q is centered. Equations for all the moments of the MMD are obtained the same way.

Since we shall later be gravimetrically mixing polymers to obtain estimates of Q/k_0, let us look at the equations relating to these mixtures. Equation [6.7], states that the MALDI-TOF MS measured total mass, G_T^{exp}, is proportional to the true mass, G_T^0:

$$G_T^0 = M_n \sum n_i \qquad [6.15]$$

$$G_T^{exp} = k_o G_T^0 (1 + Q/k_o (M_w^0 - M_0)) \qquad [6.16]$$

Consider now a mixture of the chemically identical polymers with end groups having different masses, or two different average molecular mass polymers, such that they cane be distinguish as separate series in the mass spectrum. Call them polymer A and polymer B. Then the measured ratio of the masses of each is:

$$\frac{G_{TA}^{exp}}{G_{TB}^{exp}} = \frac{k_{oA} G_{TA}^o}{k_{oB} G_{TB}^o} \left\{ \frac{1 + (Q_A/k_{oA})(M_{WA}^o - M_o)}{1 + (Q_B/k_{oB})(M_{WB}^o - M_o)} \right\} \qquad [6.17]$$

Notice we perform the expansions for both polymer distributions A and B around the same M_o. By simple algebra we obtain:

$$\frac{G_{TA}^{exp}}{G_{TB}^{exp}} = \frac{k_{oA} G_{TA}^o}{k_{oB} G_{TB}^o} \left\{ \frac{1 + (Q_A/k_{oA})(M_{WA}^o - M_{WB}^o)}{1 + (Q_A/k_{oA})(M_{WB}^o - M^o)} \right\} \qquad [6.18]$$

where we note Q_A, Q_B, k_{oA} and k_{oB} are all functions of M^o. Notice two things are needed for the ratios of the true masses to be obtained from the measured ratio of the MALDI-TOF MS. First, the ratio k_{oA}/k_{oB} needs to be determined. If polymers A and B are chemically the identical with different average molecular masses then the ratio is one because we have taken the expansion point for the Taylor's expansion as M_o and that was chosen to be the same for both. If the repeat group is identical but the end group is different then it is possible the ionization probability may be different. In this paper we have chosen end groups with similar masses (58 u to 114 u) and with the same, minimal polarity (*n*-butyl and *n*-octyl) so that ablation and ionization should be controlled by the main polymer PS repeat. (Since we expect little silver charging on the butyl or octyl end group, corrections should be much less than [number of repeats units]$^{-1}$). Thus we may assume the ratio of $k_{oA}/k_{oB}=1$. The same argument applies to Q, i.e. that it is independent of the end group and only depends on M_0 and the other machine parameters.

Thus we have from equation [6.18]

$$\frac{G_{TA}^{exp}}{G_{TB}^{exp}} = \frac{G_{TA}^o}{G_{TB}^o}\left\{\frac{1+(Q/k_o)(M_{WA}^o - M_{WB}^o)}{1+(Q_A/k_{oA})(M_{WB}^o - M^o)}\right\} \quad [6.19]$$

Simple algebra leads us to:

$$\frac{G_{TA}^{exp}}{G_{TB}^{exp}} = \frac{G_{TA}^0}{G_{TB}^0}\left\{1 + \frac{(Q/k_0)(M_{wA}^0 - M_{wB}^0)}{1+(Q/k_0)(M_{wB}^0 - M_0)}\right\} \quad [6.20]$$

We shall later plot $\dfrac{G_{TA}^{exp}}{G_{TB}^{exp}}$ versus $\dfrac{G_{TA}^0}{G_{TB}^0}$. The slope, $\dfrac{G_{TA}^{exp}/G_{TB}^{exp}}{G_{TA}^0/G_{TB}^0}$, of that plot is:

$$\frac{G_{TA}^{exp}}{G_{TB}^{exp}} \bigg/ \frac{G_{TA}^{0}}{G_{TB}^{0}} = \left\{ 1 + \frac{(Q/k_0)(M_{wA}^{0} - M_{wB}^{0})}{1 + (Q/k_0)(M_{wB}^{0} - M_0)} \right\} \qquad [6.21]$$

As before with equation [2.19] the reader should notice that if M_{wB}^{0} is close to M_0 the term $(Q/k_o)(M_{wB}^{0} - M_0)$ is small compared to 1 which means the slope largely depends on the difference $(M_{wA}^{0} - M_{wB}^{0})$ and on the ratio (Q/k_0). This concept will be used later on in the data analysis to obtain estimates of (Q/k_0) in a self-consistent, iterative manner.

We close this section saying that the gravimetric calibration of the signal axis using chemically identical polymers can avoid the issues pertaining to the uncertainties arising from ablation, ionization, and detection. However, uncertainties in repeatability and consistency in sample preparation as well as in data analysis still affect the gravimetric calibration technique. Our earlier work in sample preparation methods [1] and data analysis [3] that were employed here were aimed at reducing these effects.

We now return to equation [6.10]. To simplify the development we can assume $M_o = M^0{}_n$. That is, we expand around the (unknown) true number average molecular mass. By doing so we obtain:

$$M_n^{exp} = M_n^0 (1 + M_n^0 (Q/k_o)(PD^0 - 1)) \qquad [6.22]$$

where M_n^{exp} is the experimentally measured M_n from MALDI.
One step further and we obtain:

$$\frac{(M_n^{exp} - M_n^0)}{M_n^0} = M_n^0 (Q/k_o)(PD^0 - 1) \qquad [6.23]$$

We close this section saying that the gravimetric calibration of the signal axis using chemically identical polymers can avoid the issues discussed in Section 4 on the uncertainties arising from ablation, ionization, and detection. However, the issues arising from uncertainties in repeatable sample preparation, and data analysis, still affect the gravimetric calibration techniques. In sample preparation, the problem of phase separation as a function of molecular mass (each component of the blend is usually of a different molecular mass) appears. In solution, the solubility of some polymers is very molecular-mass dependent (e.g., polyethylene glycol solubility in tetrahydrofuran). As the solvent evaporates in the dried droplet technique the oligomers of different molecular masses come out of solution at different rates and thus may find themselves at different matrix:polymer ratios. In data analysis the issues of determining the baseline consistently and drawing areas consistently is still an issue and may incorrectly represent the amount of mass in one or the other polymer in the blend. Further, if one polymer is a small fraction of the other, the experimental noise will cause major problems with peak integration.

7. Experimental Method for the Calibration of the Signal Axis

We wish to look at the calibration of the signal axis using the expression derived in section 6. Calibration of the signal axis has been done before, specifically by Shimada, *et al.* [10, 11] by fractionating the polymer into distinct oligomers with supercritical fluid chromatography (SFC) and then using these materials to calibrate the spectrometer and obtain the MMD of the whole polymer from these calibrations. Our method does not depend on the SFC fractionation but rather uses narrow MMD polymers without fractionation with different end groups so they can easily be distinguished in the mass spectrum.

7.1 Samples and Reagents

Three octyl-initiated (M_n of 6200 u, 9180 u, and 12 600 u (as measured by the vendor using SEC with multiple-angle laser light scattering) and one butyl-initiated (M_n of 9100 u) polystyrene samples were synthesized by Scientific Polymer Products, Inc. (Ontario, NY). These four polymers will be referred to as OctylB, OctylA, OctylC, and ButylA, respectively. SRM 2888, a PS of about 6800 u made by Polymer Source (Dorval, Québec, Canada) was used as received. Two other PS with butyl end groups were also used in this experiment: PS10k (M_n of 10 000 u) from Polymer Standards Service, GmbH (Mainz, Germany), and catalog number 745 (lot 02) (M_n of 8000 u) from Scientific Polymer Standards (Ontario, NY). The matrix used in these experiments was *all-trans* retinoic acid (RA) purchased from Sigma-Aldrich (Milwaukee, WI) and used as received. The retinoic acid was stored in a freezer to preserve it. Silver trifluoroacetate (AgTFA) was purchased from Sigma-Aldrich and used as received. The solvent used for all the experiments was unstabilized tetrahydrofuran (THF) also from Sigma-Aldrich. The THF was checked before each experiment for presence of peroxides using Quantofix Peroxide 100 (Macherey-Nagel, Düren, Germany) test strips. No THF with more than 1 mg/L peroxides was used.

7.2 MALDI-TOF Mass Spectrometry

All experiments were performed on a Bruker Daltonics (Billerica, MA) Reflex II MALDI-TOF MS with a 2 GHz digitizer. The flight distance is nominally 1.5 m from source to microchannel plate detector. A Laser Science Incorporated (Franklin, MA) model LSI-337 nitrogen gas laser operating at 337 nm with an approximately 3 ns pulse width was utilized. The laser pulse energy was tested both before and after the experiments on each gravimetric composition using a Laser Probe Inc. (Utica, NY) universal radiometer (model RM-3700) with a model RJP-465 energy probe. Twenty events were averaged to compensate for the inherent shot-to-shot variation found in nitrogen gas lasers. All mass spectra in this experiment were obtained in reflectron mode. Mass axis calibration was performed as described in Section 5.

7.3 Instrument Optimization for Signal Axis

The instrument was tuned for quantitation by numerical optimization using a stochastic gradient approximation method to find a region in instrument parameter space with the least mass bias. This is described in full detail in a previous publication [2]. We briefly describe the method here so we may refer to it in our discussion of Type B uncertainties for the SRM 2881.

A new approach was used for the selection of optimal instrument parameters that yield a mass spectrum which best replicates the molecular mass distribution of a synthetic polymer. An implicit filtering algorithm was shown to be a viable method to find the best instrument settings while simultaneously minimizing the total number of experiments that need to be performed. This includes considerations of when to halt the iterative optimization process at a point when statistically-significant gains can no longer be expected. An algorithm to determine the confidence intervals for each parameter is also given. To do this we used a mixture of ButylA, OctylB, and OctylC in a mass ratio of 10:70:20 respectively. Figure 1 shows one of these spectra. Laser intensity, extraction voltage, extraction delay time, lens voltage, and detector voltage were each varied in order to find their optimal instrument values for measuring the correct molecular mass distribution of the polymer mixture. The sample plate voltage and reflectron voltages were held constant because their numerical relationship is determined by the geometry of the ion optics. The function minimized to determine the optimum instrument parameters was termed $J(x)$ where x is a vector of all the machine parameters:

$$J(x) = \left(\frac{G_{TOB}^{exp}}{G_{TBA}^{exp}} - \frac{G_{TOB}^{0}}{G_{TBA}^{0}} \right)^2 + \left(\frac{G_{TOC}^{exp}}{G_{TBA}^{exp}} - \frac{G_{TOC}^{0}}{G_{TBA}^{0}} \right)^2 \qquad [7.1]$$

where, similar to equation [6.16], G_{TOB}^{exp} is the MALDI computed total mass of OctylB (OB) and G_{TOB}^{0} is the gravimetric mass of OctylB in the mixture and equivalently OC in the subscripts refer to OctylC and BA in the subscripts refer to ButylA. When $J(x)=0$ the integrated mass spectral peak ratios equal the gravimetric ratios and the instrument settings are at their optimum values. $J(x)$ was minimized to a small, but non-zero value.[2] Later, in the discussion of uncertainty in section 5.2, it will be noted that the derivatives of $J(x)$ at the function minimum play a critical role in determining type B uncertainties for the instrument parameters.

7.4 Sample Preparation

Ten mg of a butyl-initiated PS and 10 mg of an octyl-initiated PS were measured on a calibrated scale, diluted in 4 mL of unstabilized THF, and used immediately. The polymer that was to be varied in the gravimetric mixture also contained 10 mg AgTFA.

The matrix was *all-trans* retinoic acid (RA). 25 mg RA in 0.5 mL unstabilized THF was placed in each of 22 bottles. Eleven bottles were labeled SA through SK designating samples A thru K. The other eleven bottles were marked BA through BK signifying blank samples containing no polymer. The polystyrene and AgTFA solutions were made at the beginning of the first day of each two-day experiment, while the matrix solutions were made fresh daily. One-half mL of THF was added to SA and BA. The butyl amount of 60 μL was added to SA along with 60 μL of AgTFA. Just 60 μL of AgTFA was added to BA. Sample A and sample G never contained the octyl-initiated polymer. All samples were deposited onto the MALDI sample target by electrospray (ES) to increase signal repeatability and reduce "hot spots" following the work of Hanton, *et al.* [1,22] We found electrospray sample preparation essential for experimental stability and reproducibility. The samples were electrosprayed using a voltage of 5 kV and a flow rate of 5 μL/min. The target was placed (4.0 ± 0.2) cm from the tip of the needle. Electrosprayed films on the order of several micrometers thick were made to avoid depletion of the target during the taking of the spectra. The target was divided in half by placing double-sided sticky tape down the center and placing Teflon film across one side. The right half of the target was electrosprayed with RA and AgTFA only (no polymer) to be used in the creation of background spectra. Reversing the film, the left half of the target was electrosprayed with polystyrene mixture, RA, and AgTFA in a 6:50:6 ratio. Each mass spectrum was the sum of 500 laser shots. Five repeats each on both the "blank" (no polymer) and the sample side of the target and was taken from randomized positions on the target.

When finished the THF was added to BA and BB which sat on the counter while Sample A was run. Then according to the scheme 60 μL of butyl and 60 μL of AgTFA was added to SB. The amount of octyl that already contained AgTFA was varied (e.g., 40 μL for a total of 100 μL of polystyrene). The total amount (100 in this case) of AgTFA was added to BB. We then proceeded as above and divided the target, spraying ½ with polystyrene mix and half with just matrix and AgTFA. This process was continually repeated running sample A thru F. The remaining jars containing RA only, along with the polystyrene and AgTFA mixes were stored overnight in the refrigerator. This process was continued on day two running samples G thru K.

Concentration of total polymer to retinoic was held about at the three-quarter point in the range of linearity found by Goldschmidt and Guttman.[18] This range was re-checked in preliminary experiments. In addition, the silver triflouroacetate concentration from the above recipe was found to work well for the linear range as long as there was a molar excess of silver. This was checked by making measurements at concentrations of silver triflouroacetate ranging from 25 % to 200 % of that value using the 10:70:20 of OctylB:ButylA:OctylC gravimetric mixture employed in the instrument optimization experiments [2] and described in section 3.4. Within the noise limits of the repeatability of the experiment there was no change of ratio of OctylB/Butyl A or OctylC/ButylA either as a function of silver triflouroacetate concentration or as a function of the time the experiment was done.

8. Data Analysis Methods

The data for one set of linearity experiments require up to two days to collect yielding 35 sets of data to analyze. Without automated data analysis methods this can lead to a severe bottleneck in standards production. The automated data analysis program must identify the peaks (as many as 80 or 100 for a mixture of two narrow distribution polymers), determine the area under each peak, and for these experiments sort the peaks into series which are identifiable by end group of polymer, sum the total area attributable to each series, determine the MMD and the MMD moments of each series and compute the apparent gravimetric mass attributable to each series. Since there are five repeats at each point the programs must also be able to handle these and any other required statistics automatically.

The transformation of the data time-of-flight to mass was done using the Bruker Daltonics *Xmass* program (version 5.1.0). Peak picking and peak integration were performed using the *MassSpectator* FORTRAN computer code [3-5] developed at NIST for standards production. *MassSpectator* makes no assumptions about peak shape and requires no smoothing or preprocessing of the data, a process which has been shown to distort peak area.[3] The algorithm is based on a time-series segmentation routine that reduces the data set to groups of three strategic points where each group defines the beginning, center, and ending of each peak located. The peak areas are found from the strategic points using a commonplace polygonal area calculation routine. Peaks with statistically insignificant height or area are then discarded. Unbiased peak integration is a cornerstone of SRM 2881.

The output files containing peak mass and peak area created by *MassSpectator* were transferred to Microsoft (Redwood, WA) *Excel* (2003 version). A spreadsheet was used since the amount of data from many integrated data sets easily fit onto a single worksheet. *Excel* has the advantage that macro scripts can be written in Visual Basic (VBA). This means that once the spreadsheet layout and input/output structure are determined all statistical software can be written in VBA. Thus, a simple VBA program was written which transfers the data and its relevant information into the spreadsheet from the output of *MassSpectator*. A second, more complex VBA program was written to recalibrate as required the mass axis of the data using the expected largest five to ten peaks from the polymer of the mix. The data were then separated into series by the polymer end group and the appropriate numbers such as total mass, total area, M_n, M_w, and M_z of each series were computed. It should be noted that even the detailed instructions for the experiment were developed and contained in the *Excel* spreadsheet. This was done so that both the experimental protocols with any modification made by the instrument operator and the data analysis of that experiment are contained together in the same file for convenience and for data record keeping. While the overall time to collect 35 spectra was 1.5 d to 2.5 d, the *MassSpectator* and *Excel* VBA software suite allowed us to analyze 35 or so spectra in (2 to 3) h.

9. Experimental Results

9.1 Calibration of Spectra—Fitting of the Data

We will now describe experimental tests of the various relationships derived in Section 6 culminating in Type A and Type B uncertainties for the MMD. For a fixed mass of matrix, as the amount of polymer increases so should the mass spectral signal from the polymer. This has been studied by Goldschmidt and Guttman for PS about 3000 u [23]. They show a relatively large range of linearity. We have repeated the same work for three polystyrenes in the range from 6000 u to 12 000 u. An example of those data is shown in Figure 2. The problem with this data is not lack of linearity but poor repeatability. The data is essentially linear but getting a fairly accurate slope seems to be particularly difficult. Extensive experience in this lab showed that fairly good repeatability using the electrospray sample preparation technique could be obtained. However, at times, even while taking great precautions, repeatability would be lost. The uncontrolled variable could not be ascertained.

Although it was necessary to show linearity in the concentration versus signal intensity relationship for single analytes, this was not sufficient to demonstrate that the proper MMD was necessarily derived from the MMS. To do this a mixture of two polymers —one of which is held at a constant polymer to matrix ratio and another that is varied needed to be measured. In this way the mixture can be used to control the polydispersity. We start with polymers having two different end groups but with similar molecular mass distributions, specifically OctylA and ButylA. OctylA was held at a fixed concentration of 25 mg polymer/ 1 g matrix while ButylA was varied from 0 mg polymer/ 1 g matrix to 50 mg polymer/ 1 g matrix. In other cases where we had Octyl end grouped polymer of 6000 u and 12000 u the butyl was held fixed at about the same matrix to polymer ratio as the Octyl A and the Octyl ended PS was varied. This is shown in Figure 4 and an example MALDI spectrum is shown in Figure 3. The fact that the slope is very close to one is indicative of the negligible effect of the different end group on the ablation or charging of the polymer. To check this in more detail at a fixed total polymer to matrix mass ratio, the octyl to butyl ratio was varied and the slope of the ratios of the two sets of data is again close to one. By repeating this method with various combinations of octyl and butyl initiated polymers the value of Q/k_0 was determined. Some examples of the data and the resulting plots are given in Figures 5 through 8.

From equations [2.23] through [2.25] the slopes $\dfrac{G_{TA}^{exp}/G_{TB}^{exp}}{G_{TA}^0/G_{TB}^0}$ are expected to be a function of molecular mass difference of the polymers, $(M_{wA}^0 - M_{wB}^0)$. This dependence of the experimental slopes on $(M_{wA}^{exp} - M_{wB}^{exp})$ is shown in Figure 9. We have plotted the slopes versus the difference in the experimental values, M_{wA}^{exp} and M_{wB}^{exp}, instead of the difference in the true values, M_{wA}^0 and M_{wB}^0, demanded by equation [6.21]. This is the first step in a iterative procedure where for the values of M_w^0 in equation [6.21] we took $Q/k_0 = 0$ initially in the term $1 + (Q/k_0)(M_{wi}^0 - M_0)$ and $1 + (Q/k_0)(M_{zi}^0 - M_0)$ for equation [6.14]. With the computed value of Q/k_0 from Figure 9 we got a new value for the terms $1 + (Q/k_0)(M_{wB}^0 - M_0)$ and for M_{wA}^0 and M_{wB}^0. A new Q/k_0 was then computed and this was carried on until an unchanging value of Q/k_0 was found. This took about five iterations. Over those iterations the value of the slope went from 6.37×10^{-5} to 6.47×10^{-5} which was the final value for Q/k_0 with an estimated type A uncertainty of 40 %.

Approximately fifteen or so Q/k_0 slopes were collected over a span of about 12 months. Stable control of sample preparation and of instrument parameters was a challenge this many experiments and time span. Some data were rejected because it was too difficult to determine the slope from the noise in the data but most runs were deemed acceptable.

10. Discussion of Type A Uncertainties

There are two Type A uncertainties that affect the MMD. The Type A uncertainty from Q/k_0, the MMS to MMD conversion factor that was estimated in section 9.1, and the Type A uncertainty from the uncertainty of the intensity of each individual oligomer in the MMS:

$$\beta_i = S_i / \sum_i S_i \qquad [9.1]$$

where β_i is the fraction of the signal for the ith oligomer with mass, m_i, from the mass spectrum. In the uncertainty analysis averages of each β_i for approximately 500 spectra of which OctylA (the polymer that became SRM 2881) was one of the two polymers mixed, or when OctylA was used alone. None of the spectra used showed any variation with M_n or M_w as a function of concentration of one or the other polymer.

10.1 Type A Uncertainties in the MALDI-TOF Mass Spectrometry Measurement

There are two sources of type A uncertainty that affect the MMD. The uncertainty from Q/k_0, the mass spectrum to MMD conversion factor, that was estimated in the previous section, and the uncertainty of the intensity of each individual oligomer in the mass spectrum:

$$\beta_i = S_i / \sum_i S_i \qquad [5.1]$$

where β_i is the fraction of the signal for the ith oligomer with mass, m_i. In the uncertainty analysis averages of each β_i for approximately 500 spectra of which OctylA (the polymer that became SRM 2881) was one of the two polymers mixed, or when OctylA was used alone were calculated.

An analysis of variance (ANOVA)[5] was done on each oligomeric β_i assuming they were independent of each other. There are two levels of the experimental data to look at the ANOVA. One is from each two day period when a given pair of polymers was run to estimate slope of the $\dfrac{G_{TA}^{exp}}{G_{TB}^{exp}}$ versus $\dfrac{G_{TA}^{o}}{G_{TB}^{o}}$ plot. This has been called by us a Polymer-Pair-run. In those runs in which Octyl A was used as one of the polymers, we are able to estimate the ANOVA for each of the oligomeric β_i of Octyl A. None of the spectra used in this analysis showed any variation in M_n or M_w as a function of concentration of one or the other constituent polymer or as a function of the time when the spectra was taken (indicating that the sample preparation solutions did not change during the Polymer-Pair-run). As stated before, the data were checked for concentration dependence of the moments M_n and M_w of the MMD. The concentration was randomized for the Polymer-Pair-run collection period with 2 or 3 repeats of concentration to check for consistency. The data was analyzed as a function of time they were run in the day and/or the order of running. None of these showed any significant trend in the moments. This also suggested the MMD was not changing significantly as a function of the overall concentration of polymer in the matrix or of the order of the running the samples.

An analysis of variance (ANOVA)[5] was done on each oligomeric β_i assuming they were independent of each other. For these at each concentration ratio five spectra were taken of the mixture on each sample preparation—making up of solution with syringes and electrospraying the solution of a ratio of two polymers with matrix and appropriate Ag salt. For these spectra it was found that the sample-to-sample variance for the fraction of a given oligomeric mass is much greater than the within-sample variance where this case within-sample refers to the five repeats of one spraying and between sample refers to different sprayings and different gravimetric mixtures. Thus, the *computed F* values generally ran much higher than those generally expected from the F computed for the each oligomeric species in distribution for $\alpha = 0.05$ with the appropriate degrees of freedom. This computed F is normally called $F_{critical}$. Assuming the oligomer β_i are independent one would expect 5% of the computed F's to be above $F_{critical}$. Often as many as 50% of the F's were above $F_{critical}$ suggesting that there is a significant variation between each sample preparation.

The second level of experiment at which we can look at the ANOVA is the individual oligomeric OctylA β_i repeatability amongst Octyl A samples run in different Polymer-Pair-runs. For each Polymer-Pair-run about 6 to 15 sample preparations are made from the same solution of each of the two polymers, silver and matrix. The average of each β_i for each oligomer of the Octyl A is obtained. The sum of within variances of all the Polymer-Pair-run can then be compared to the variances between all Polymer-Pair-run used in the average β_i used to certify the sample. Assuming the oligomer β_i are independent one would expect 5% of the computed F's to be above $F_{critical}$. Again often as many as 50% of the F's were above $F_{critical}$ suggesting that there is a significant variation between each sample preparation.

Two other methods were used to check the reproducibility of the data. First, linearity of concentration versus signal intensity was found for all data sets. Second, all data sets had a few (at least two) sets of data points in which the concentration ratios of the two polymers were repeated. In all these the noise within one repeat of five data sets was larger than the repeatability between the solutions made up the same—indicating the repeatability of our making of the solutions was excellent. This showed that the sample preparation method used within one set of experiments was good.

The data was analyzed in two additional ways to check for self-consistency. In the first, a single peak was given a fixed value (the average value for that peak of all spectra) and the rest of the peaks the spectrum under consideration were scaled accordingly. This did not change the overall distribution of values of the *p*-value computed indicating that spectrum-to-spectrum differences were a leading cause of type A uncertainty. In the second, four adjacent data points in a spectrum were summed to one reducing the number of points per peak from 64 points to 16 points. Assuming the oligomer β_i obtained with this course graining of the spectrum (or smoothing of the spectrum) are independent one would expect 5% of the computed F's to be above $F_{critical}$ obtained for these data. Again often as many as 50% of the F's were above $F_{critical}$ suggesting that there is a significant variation between each sample preparation.

11. Discussion of Type B Uncertainties

11.1 Instrument contribution to Type B uncertainty

In an earlier paper [2] a novel approach is described for the selection of optimal instrument parameters that yield a mass spectrum which best replicates the molecular mass distribution of a synthetic polymer. The application of stochastic gradient approximation algorithms was shown to be a viable method to find the best instrument settings while simultaneously minimizing the total number of experiments that need to be performed. This includes considerations of when to halt the iterative optimization process at a point when statistically-significant gains can no longer be expected. An algorithm to determine the confidence intervals for each parameter is also given. Details on sample preparation and data analysis that ensure stability of the measurement over the time scale of the optimization experiments are provided.

We can write $J(x)$ from equation [7.1] in terms of the calibration equation for the signal axis:

$$J(x) = \left(\frac{G^0_{TOB}}{G^0_{TBA}}\right)^2 \left[\frac{G^{exp}_{TOB}/G^{exp}_{TBA}}{G^0_{TOB}/G^0_{TBA}} - 1\right]^2 + \left(\frac{G^0_{TOC}}{G^0_{TBA}}\right)^2 \left[\frac{G^{exp}_{TOC}/G^{exp}_{TBA}}{G^0_{TOC}/G^0_{TBA}} - 1\right]^2 \qquad [11.1]$$

Substituting equation [6.19] into equation [11.1] yields:

$$J(x) = (Q/k_0)^2 \left[\left(\frac{G^0_{TOB}}{G^0_{TBA}}\right)^2 \left(\frac{(M^0_{wOB} - M^0_{wBA})}{1+(Q/k_0)(M^0_{wBA} - M_0)}\right)^2 + \left(\frac{G^0_{TOC}}{G^0_{TBA}}\right)^2 \left(\frac{(M^0_{wOC} - M^0_{wBA})}{1+(Q/k_0)(M^0_{wBA} - M_0)}\right)^2\right]^2$$

[11.2]

From this we obtain partial derivatives of Q/k_0 as:

$$\frac{\delta \ln(J(x))}{\delta p} \approx 2\frac{\delta \ln(Q/k_0)}{\delta p} \qquad [11.3]$$

where p is any instrument parameter. Notice as before we have taken the term $1+(Q/k_0)(M^0_{wBA}-M_0)$ to be very near 1 and second order in all corrections to the derivative. All other terms are independent of (Q/k_0).

In Table 2 we show the individual derivatives obtained for each machine parameter and their effect on the uncertainty of the MALDI calibration parameter as found in Table 1. Inspection of Table 1 tells us that the statistical uncertainty dominates all Type B uncertainty.

We need to point out that the optimization method along with the calibration model allows us to estimate the Type B uncertainty which have been optimized by the method described in that paper. Generally this is much simpler that what had to be done previously.

11.2 Sample preparation contribution to Type A and Type B uncertainty

This section considers the estimated error involved in sample preparation, particularly the gravimetric aspects. Specifically, this involves the masses of each of the polymers in the solutions, the volumes of solution used to make up the concentration ratio, and finally the repeatability of the MALDI made from two different solutions of the apparently same concentration ratio.

11.2.1 Mass and Volume effects

Initial solutions were made from polymer samples weighing from 5 mg to 6 mg precise to 0.1 mg and added to 4 mL of THF. The weighing was found stable to 0.1 mg thus we estimate an uncertainty of $[\sqrt{2}\cdot(0.1/5.0)] \approx 3\%$. Once the stock solutions were prepared volumes were combined to create various polymer mixtures. Although pipettes were initially used to measure out solutions, these were found to poorly reproduce volumes in the μL range for THF. Instead, Hamilton glass syringes of 100 μL total volume were used and found to reproduce the volumes well. The reproducibility of the 100 μL syringe was determined to be about 3 %. Repeatability of a 5 mL glass syringe for a 4 mL volume used to make up the initial solutions of polymers was determined to be about 4 %. To avoid contamination and calibration problems separate syringes were used for each polymer throughout the experiments.

Two other methods were used to check the reproducibility of the data. First, linearity of concentration versus signal intensity was found for all data sets. Second, all data sets had a few (at least two) sets of data points in which the concentration ratios of the two polymers were repeated. In all these the noise within one repeat of five data sets (to be described later) was larger than the repeatability between the solutions made up the same—indicating the repeatability of our making of the solutions was excellent. This showed that the sample preparation method used within one set of experiments was good.

11.2.2 Type B uncertainty related to sample preparation and ablation

The repeatability of the electrospray part of the sample preparation and repeatability of the ablation was difficult to determine. Once a target was sprayed, it as placed in the instrument and 500 laser shots for the sample side of the target were taken followed by 500 laser shots of the blank side of the target. This was repeated five times. Mass peaks were then integrated and separated into their appropriated groups by end group mass. Next the ratios from these integrated peak sums taken to give the integrated ratios between the two polymers in the sample. These integrated ratios are displayed on each graph in Figures 4, 6 and 8. Uncertainty in the results can be traced to the repeatability of the MALDI measurement. Even when going to great lengths in sample preparation and by using a large number of laser shots per spectrum, the MALDI process itself proved to be highly variable. Thus we take the type B uncertainty of this process to be covered by the Type A (random) uncertainty.

12. Discussion of Certified Values

The concentration β_i for each oligomer (with associated uncertainties) for the OctylA polystyrene MMS converts it into SRM 2881 with a certified MMD. All Type B uncertainty is small compared to the two forms of Type A uncertainty, gravimetric and ablation. Here we consider the issue of why the Type B uncertainty is small compared to the Type A. In earlier SRM work using light scattering to determine the M_w or osmometry to determine the M_n, we generally were able to design the experiment so Type A uncertainty was comparable or somewhat smaller then the Type B uncertainty. Here this is certainly not the case. We may take it as lack of control over some parts of the experiment—in particular the sample preparation. The sample preparation is done in a way hoping to obtain a good sampling of the MMD of the original polymer by the use of a spraying technique where it is expected that that the rapid droplet evaporation will allow us to obtain uniformity in the sample surface. But the electrospray may easily be affected by a variety of things we did not control—humidity in room during sample preparation and subtle changes in the electrospray cone character leading to uncontrolled droplet size. Furthermore, the ablation process is a nonequilibrium event; making the chemistry of the plume is itself a variable process. Any expectations of high-precision repeatability are unlikely.

There is a second source of Type A uncertainty. There is the source from the repeatability of the fundamental data, i.e., in β_i, as discussed above. However, there is also a source of uncertainty in the correction term for Q/k_0. Table 3 gives the values of the mean value of the experimental β_i the experimental fraction of the signal in peak of mass m_i, with its statistical standard deviation (the contribution to the Type A uncertainty). It is unclear that these are independent uncertainties but having no other knowledge we must assume they are independent and add them in the usual way by using the sum of squares. (The reader is reminded that there are no contributions from the Type B uncertainties because they are so small.) Table 1 offers a listing of the contributions to the uncertainty from all sources. It is unclear that these are independent uncertainties but having no other knowledge we must assume they are independent and add them in the usual way by using the sum of squares. Table 3 offers the MMD for the polymer as certified by this method with the uncertainties of each β_i for the cumulative distribution form 1 % to 99 %. Table 4 is a reduced version of this table found on the SRM certificate. Figure 10 is a plot of the final certified values found in Table 4.

13. Conclusions

We conclude with a few final words about the Taylor's expansion mathematical model. The impetus for the model came from noticing in many earlier published reports that k_i appeared to be a slowly-varying function of mass. A Taylor's expansion that is linear in mass creates a useful computational bridge from mass spectra to measurable mass moments. For this reason we chose to build on this foundation. However, if some other variable (e.g., time in time-of-flight mass separation) were shown to have a longer range of linearity then an expansion in that variable would provide a more accurate fitting function.

Our testing of the mathematical model is far from exact but appears to be robust provided a few conditions are met. Most importantly, for a given set of instrument parameters and for a sample preparation with some modicum of homogeneity and repeatability, a fairly wide region of instrument space where the polymer-to-matrix mass ratio results in a proportional number of polymer ions arriving at the detector must exist. This proportionality constant must be a slowing varying function of mass so that it can be linearly approximated. Additionally, this region of concentration space must have a high enough polymer concentration for us to obtain a good signal-to-noise ratio. Further improvements could be made in two areas: 1) improved reproducibility in the data likely through better sample preparation methods and a fuller understanding of the MALDI process, and 2) an estimate of the functional dependence of Q on the various machine parameters. This latter improvement can only come from a mathematical description of the total instrument behavior, from ablation through mass separation and detection to signal digitization, which is currently beyond reach for MALDI-TOF mass spectrometry.

Acknowledgements

The nuclear magnetic resonance experiments were performed and interpreted by David Vanderhart of the NIST Polymers Division. John Lu of the NIST Statistical Engineering Division provided statistical expertise in close consultation with the authors.

References

1. S. D. Hanton, I. Z. Hyder, J. R. Stets, K. G. Owens, W. R. Blair, C. M. Guttman and A. A. Giuseppetti, *J. Am. Soc. Mass Spectrom.*, 2004, **15**, 168-179.

2. W. E. Wallace, C. M. Guttman, K. M. Flynn and A. J. Kearsley, *Anal. Chim. Acta*, 2007, **604**, 62-68.

3. W. E. Wallace, A. J. Kearsley and C. M. Guttman, *Anal. Chem.*, 2004, **76**, 2446-2452.

4. A. J. Kearsley, W. E. Wallace, J. Bernal and C. M. Guttman, *Appl. Math. Lett.*, 2005, **18**, 1412-1417.

5. A. J. Kearsley, *J. Res. Natl. Inst. Stand. Technol.*, 2006, **111**, 121-125.

6. N. G. Natrella, *Experimental Statistics*, National Bureau of Standards Handbook 91, US Government Printing Office, Washington, D.C., 1963.

7. L. Huber, *Hewlett-Packard*, 1989, 89-100.

8. P. A. Clay and R. G. Gilbert, *Macromolecules*, 1995, **28**, 552-556.

9. W. Lee, H. Lee, J. Cha, T. Chang, K. J. Hanley and T. P. Lodge, *Macromolecules*, 2000, **33**, 5111-5115.

10. K. Shimada, M. A. Lusenkova, K. Sato, T. Saito, S. Matsuyama, H. Nakahara and S. Kinugasa, *Rapid Comm. Mass Spectrom.*, 2001, **15**, 277-282.

11. K. Shimada, R. Nagahata, S. Kawabata, S. Matsuyama, T. Saito and S. Kinugasa, *J. Mass Spectrom.*, 2003, **38**, 948-954.

12. W. Yan, J. A. Gardella and T. D. Wood, *J. Am. Soc. Mass Spectrom.*, 2002, **13**, 914-920.

13. H. Chen, M. He, J. Pei and H. He, *Anal. Chem.*, 2003, **75**, 6531-6535.

14. H. Chen and M. He, *J. Am. Soc. Mass Spectrom.*, 2005, **16**, 100-106.

15. D. C. Schriemer and L. Li, *Anal. Chem.*, 1997, **69**, 4169-4175.

16. D. C. Schriemer and L. Li, *Anal. Chem.*, 1997, **69**, 4176-4183.

17. H. Zhu, T. Yalcin and L. Li, *J. Am. Soc. Mass Spectrom.*, 1998, **9**, 275-281.

18. R. J. Goldschmidt and C. M. Guttman, *ACS Polym. Preprints*, 2000, **41**, 647-648.

19. G. Montaudo, M. S. Montaudo, C. Puglisi and F. Samperi, *Rapid Comm. Mass Spectrom.*, 1995, **9**, 453-460.

20. C. N. McEwen, C. Jackson and B. S. Larsen, *Int. J. Mass Spectrom. Ion Proc.*, 1997, **160**, 387-394.

21. H. Sato, N. Ichieda, H. Tao and H. Ohtani, *Anal. Sci.*, 2004, **20**, 1289-1294.

22. S. D. Hanton, P. A. C. Clark and K. G. Owens, *J. Am. Soc. Mass Spectrom.*, 1999, **10**, 104-111.

23. R. J. Goldschmidt and C. M. Guttman, 47th Conference on Mass Spectrometry and Allied Topics, Dallas, TX, 1999.

24. B. N. Taylor and C. E. Kuyatt, *Guidelines for Evaluating and Expressing the Uncertainty of NIST Measurement Results* NIST Technical Note 1297, US Government Printing Office, Washington, D.C., 1994

Figures

1. MALDI-TOF mass spectra of a mixture of three narrow-polydispersity polystyrenes. Top panel shows data along with background spectrum used in data processing by *MassSpectator*. Bottom panel shows bar graph of relative ion intensities.
2. ButylA linearity of MALDI signal intensity vs. polymer concentration
3. OctylA and ButylA mixture mass spectrum
4. OctylA and ButylA concentration linearity graph
5. OctylB and ButylA mixture mass spectrum
6. OctylB and ButylA concentration linearity graph
7. OctylC and ButylA mixture mass spectrum
8. OctylC and ButylA concentration linearity graph
9. Estimation of Q/k_0 for all data
10. Plot of the final certified values for SRM 2881

Figure 1

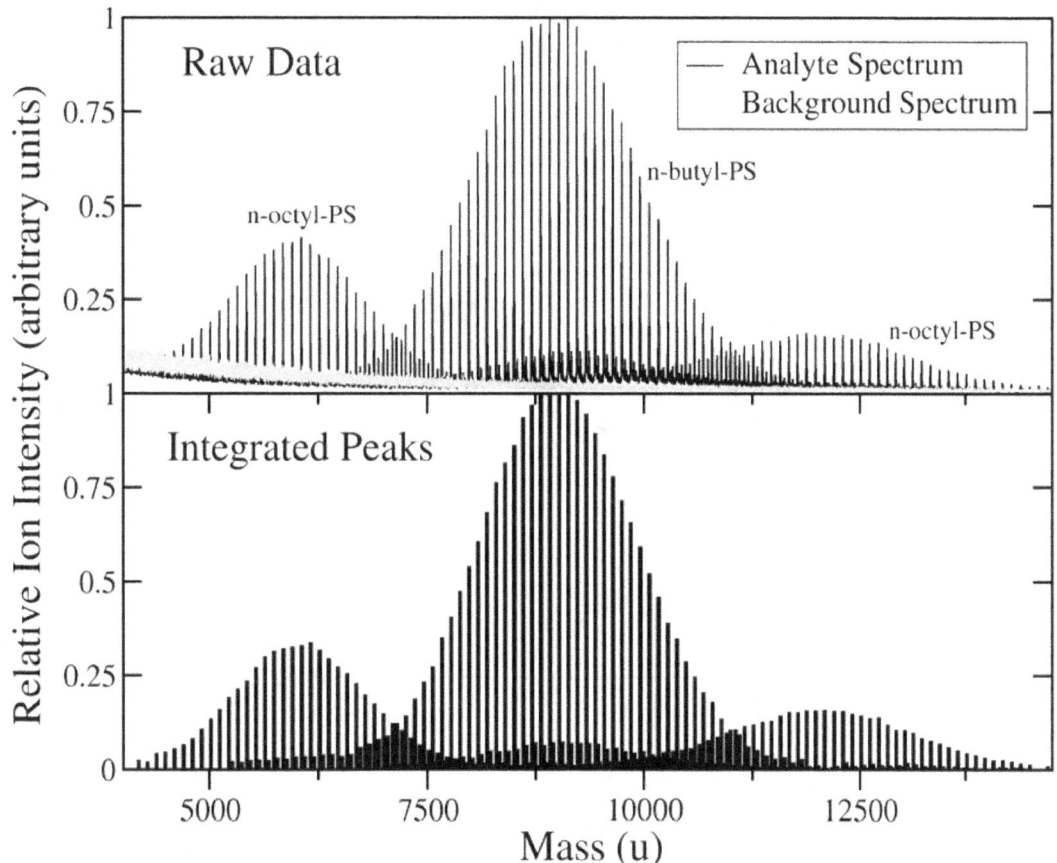

Figure 2

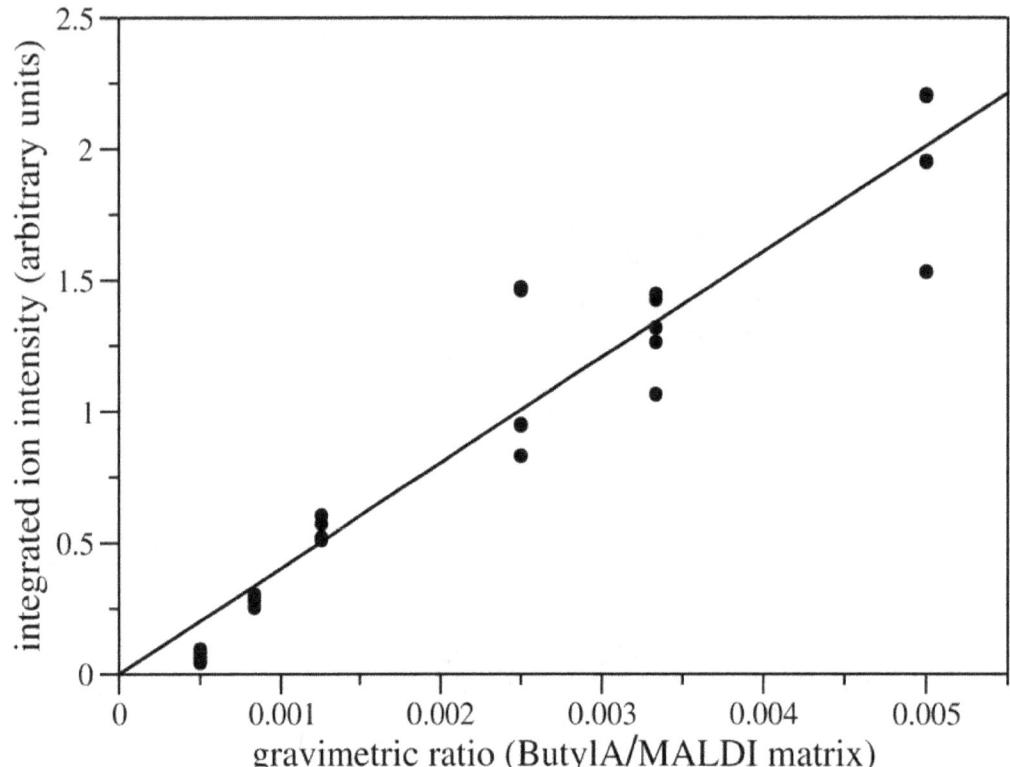

Figures 3 (inset) and 4 (main)

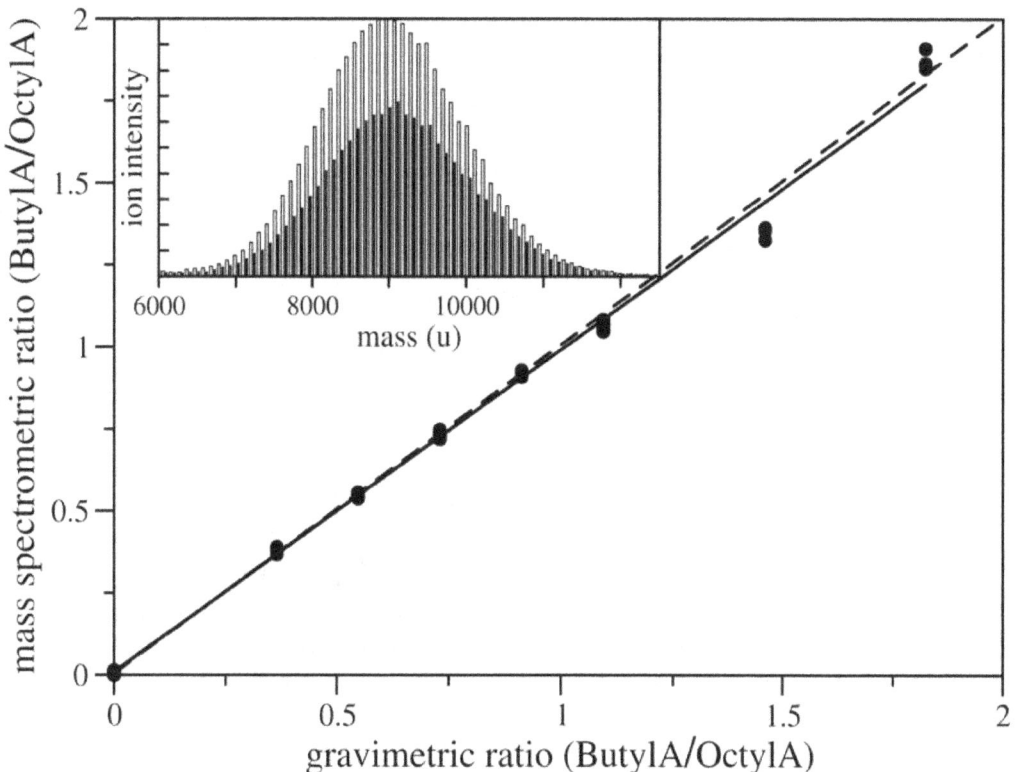

Figures 5 (inset) and 6 (main)

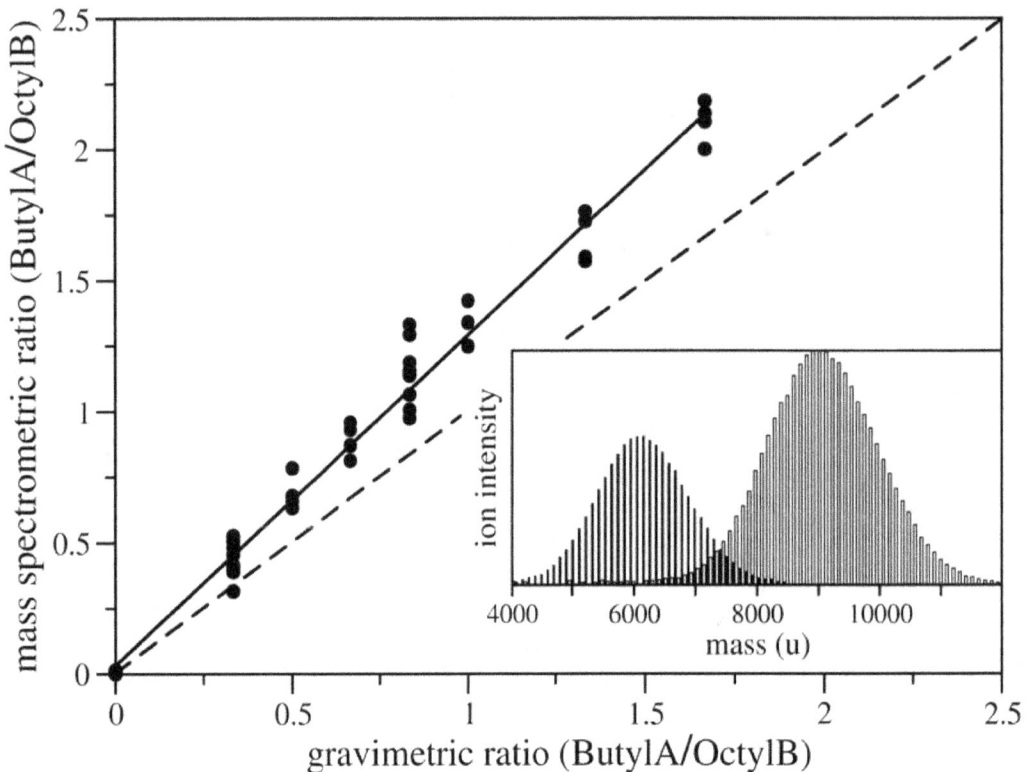

Figures 7 (inset) and 8 (main)

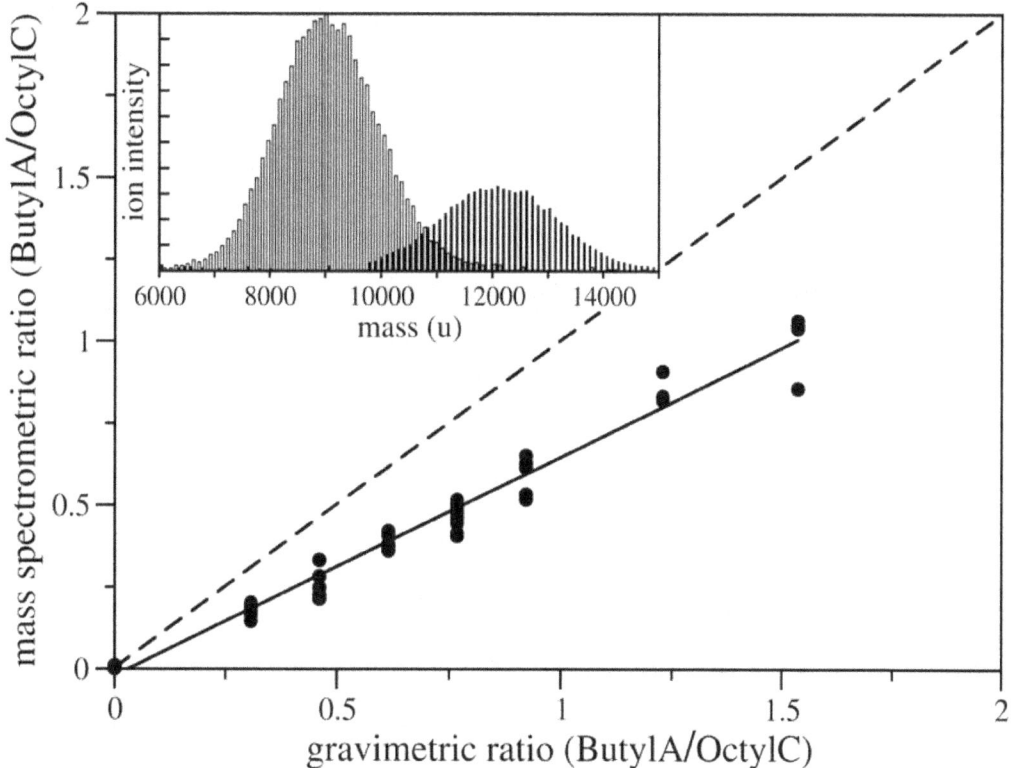

Figure 9

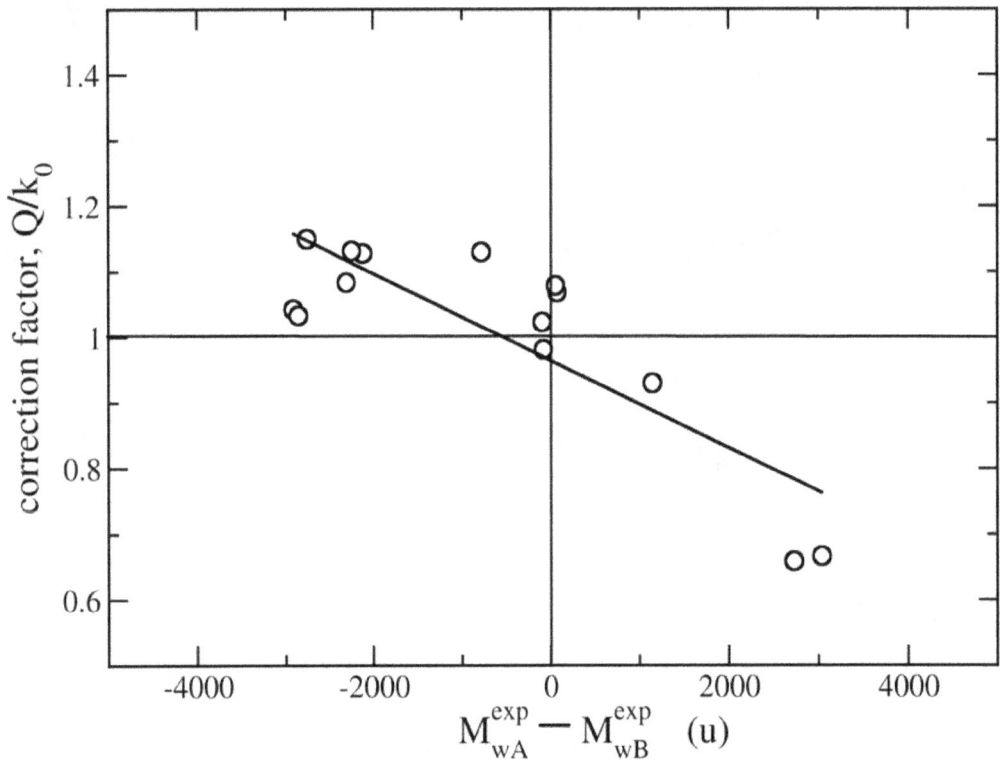

Figure 10

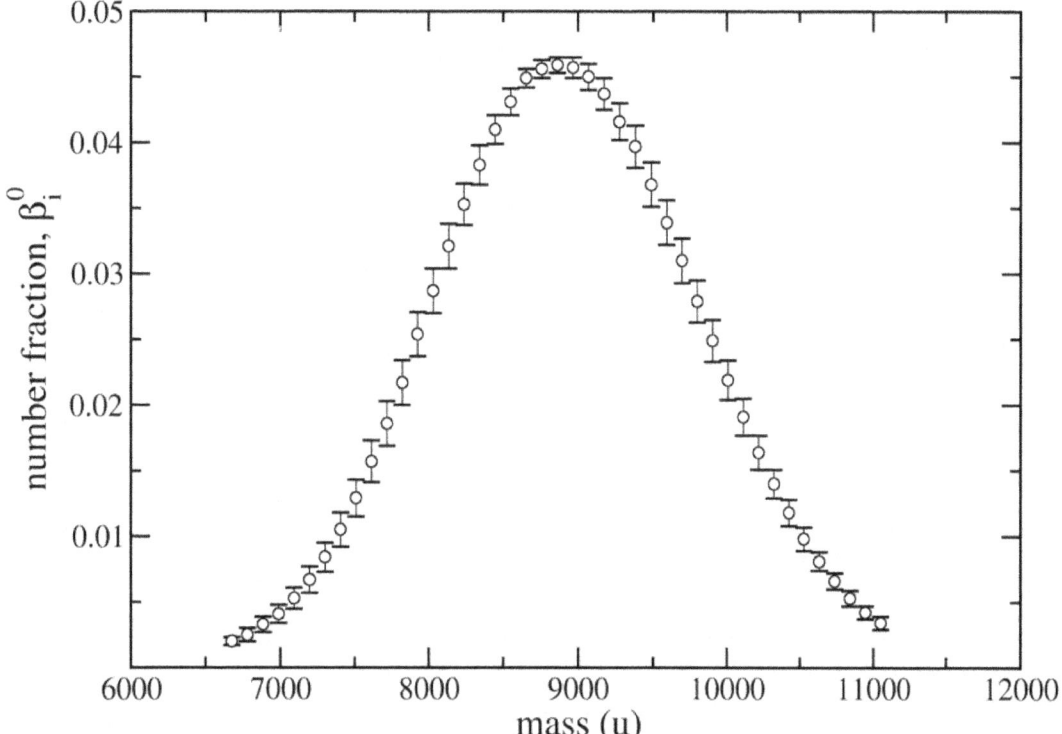

Tables

1. Q/k_0 with its estimated uncertainty
2. Individual derivatives obtained for each machine parameter and their effect on the uncertainty
3. Values of oligomer concentration at each mass of oligomer, m_i, with estimations of uncertainty from various sources
4. Certified Values of oligomer fractions for certificate

Table 1

Type A Uncertainties	All	≈ 40%
Type B Uncertainties		
Instrumentation Related Uncertainties		
	Acceleration voltage	<0.5%
	Detector voltage	0.245%
	Laser attenuation	0.15%
	Delay time	< 0.5%
	Extraction voltage	0.029%
	Lens voltage	0.014%
Sample Preparation Uncertainties		
	Weighing uncertainties	≈ 3%
	Syringe uncertainties	≈ 4%

Table 2

Instrument Parameter	Optimal Setting +/- Confidence Interval	% Type B Uncertainty Contribution to (Q/ko)
Detector Voltage	1.7 +/- 0.03 kV	0.245 %
Laser Intensity	1.86 +/- 0.11 mJ/pulse	0.15 %
Delay Time	500 ns	—
Extraction Voltage	18.2 +/- 0.80 kV	0.029 %
Lens Voltage	8.6 +/- 2.0 kV	0.014 %

Table 3

Repeat number, i	Oligomer mass (g/mole)	number fraction, β_i^0	standard deviation (%) for β_i^0	% correction for β_i^0 from Q/k_0	total Type A and Type B uncertainty in β_i^0 (footnote 1)	standard uncertainty on mean, β_i^0	% overall uncertainty of β_i^0 (footnote 2)	Uncertainty of Fraction, β_i^0	Cumulative MMD
62	6570.68	0.0018	38.90 %	-19.44 %	39.67 %	13.22 %	26.45 %	0.0005	1.03 %
63	6674.82	0.0020	40.25 %	-18.60 %	40.93 %	13.64 %	27.29 %	0.0005	1.23 %
64	6778.96	0.0025	43.42 %	-17.77 %	44.00 %	14.67 %	29.33 %	0.0007	1.48 %
65	6883.1	0.0031	32.80 %	-16.94 %	33.49 %	11.16 %	22.33 %	0.0007	1.79 %
66	6987.24	0.0041	32.21 %	-16.11 %	32.85 %	10.95 %	21.90 %	0.0009	2.20 %
67	7091.38	0.0048	30.62 %	-15.27 %	31.23 %	10.41 %	20.82 %	0.0010	2.68 %
68	7195.52	0.0063	31.14 %	-14.44 %	31.67 %	10.56 %	21.11 %	0.0013	3.30 %
69	7299.66	0.0078	25.63 %	-13.61 %	26.20 %	8.73 %	17.47 %	0.0014	4.08 %
70	7403.8	0.0090	23.52 %	-12.77 %	24.07 %	8.02 %	16.05 %	0.0014	4.98 %
71	7507.94	0.0116	19.95 %	-11.94 %	20.51 %	6.84 %	13.67 %	0.0016	6.13 %
72	7612.08	0.0135	19.99 %	-11.11 %	20.48 %	6.83 %	13.65 %	0.0018	7.48 %
73	7716.22	0.0163	16.52 %	-10.27 %	17.02 %	5.67 %	11.35 %	0.0019	9.12 %
74	7820.36	0.0191	14.39 %	-9.44 %	14.88 %	4.96 %	9.92 %	0.0019	11.03 %
75	7924.5	0.0223	11.30 %	-8.61 %	11.81 %	3.94 %	7.88 %	0.0018	13.26 %
76	8028.64	0.0262	10.26 %	-7.77 %	10.72 %	3.57 %	7.15 %	0.0019	15.88 %
77	8132.78	0.0296	8.34 %	-6.94 %	8.79 %	2.93 %	5.86 %	0.0017	18.84 %
78	8236.92	0.0322	7.52 %	-6.11 %	7.91 %	2.64 %	5.27 %	0.0017	22.06 %
79	8341.06	0.0353	4.95 %	-5.28 %	5.38 %	1.79 %	3.59 %	0.0013	25.59 %
80	8445.2	0.0385	3.92 %	-4.44 %	4.30 %	1.43 %	2.87 %	0.0011	29.44 %
81	8549.34	0.0404	4.85 %	-3.61 %	5.06 %	1.69 %	3.37 %	0.0014	33.48 %
82	8653.48	0.0423	4.14 %	-2.78 %	4.29 %	1.43 %	2.86 %	0.0012	37.72 %
83	8757.62	0.0440	4.09 %	-1.94 %	4.17 %	1.39 %	2.78 %	0.0012	42.12 %
84	8861.76	0.0446	4.18 %	-1.11 %	4.20 %	1.40 %	2.80 %	0.0013	46.58 %
85	8965.9	0.0464	5.49 %	-0.28 %	5.49 %	1.83 %	3.66 %	0.0017	51.22 %
86	9070.04	0.0446	5.89 %	0.56 %	5.89 %	1.96 %	3.93 %	0.0018	55.68 %
87	9174.18	0.0440	6.38 %	1.39 %	6.40 %	2.13 %	4.27 %	0.0019	60.07 %
88	9278.32	0.0424	6.49 %	2.22 %	6.55 %	2.18 %	4.37 %	0.0019	64.31 %
89	9382.46	0.0404	6.92 %	3.06 %	7.02 %	2.34 %	4.68 %	0.0019	68.35 %
90	9486.6	0.0380	7.13 %	3.89 %	7.30 %	2.43 %	4.86 %	0.0018	72.15 %
91	9590.74	0.0359	7.05 %	4.72 %	7.30 %	2.43 %	4.87 %	0.0017	75.73 %
92	9694.88	0.0325	7.91 %	5.55 %	8.22 %	2.74 %	5.48 %	0.0018	78.98 %
93	9799.02	0.0297	8.23 %	6.39 %	8.62 %	2.87 %	5.75 %	0.0017	81.96 %
94	9903.16	0.0268	8.45 %	7.22 %	8.93 %	2.98 %	5.95 %	0.0016	84.64 %
95	10007.3	0.0242	8.47 %	8.05 %	9.07 %	3.02 %	6.04 %	0.0015	87.06 %
96	10111.44	0.0207	9.99 %	8.89 %	10.60 %	3.53 %	7.07 %	0.0015	89.13 %
97	10215.58	0.0182	10.38 %	9.72 %	11.08 %	3.69 %	7.39 %	0.0013	90.95 %
98	10319.72	0.0155	8.24 %	10.55 %	9.26 %	3.09 %	6.17 %	0.0010	92.50 %
99	10423.86	0.0139	14.47 %	11.39 %	15.17 %	5.06 %	10.12 %	0.0014	93.89 %
100	10528	0.0113	9.94 %	12.22 %	11.07 %	3.69 %	7.38 %	0.0008	95.02 %
101	10632.14	0.0096	11.89 %	13.05 %	12.99 %	4.33 %	8.66 %	0.0008	95.98 %
102	10736.28	0.0077	7.35 %	13.89 %	9.21 %	3.07 %	6.14 %	0.0005	96.76 %

103	10840.42	0.0063	8.93 %	14.72 %	10.69 %	3.56 %	7.13 %	0.0005	97.39 %
104	10944.56	0.0051	10.61 %	15.55 %	12.30 %	4.10 %	8.20 %	0.0004	97.90 %
105	11048.7	0.0042	17.30 %	16.38 %	18.50 %	6.17 %	12.33 %	0.0005	98.32 %
106	11152.84	0.0031	9.57 %	17.22 %	11.79 %	3.93 %	7.86 %	0.0002	98.63 %
107	11256.98	0.0026	23.54 %	18.05 %	24.62 %	8.21 %	16.42 %	0.0004	98.89 %
108	11361.12	0.0019	16.90 %	18.88 %	18.51 %	6.17 %	12.34 %	0.0002	99.09 %
109	11465.26	0.0020	88.14 %	19.72 %	88.49 %	29.50 %	59.00 %	0.0012	99.28 %
110	11569.4	0.0013	49.49 %	20.55 %	50.16 %	16.72 %	33.44 %	0.0004	99.41 %
111	11673.54	0.0011	58.92 %	21.38 %	59.53 %	19.84 %	39.69 %	0.0004	99.52 %
112	11777.68	0.0009	72.09 %	22.22 %	72.64 %	24.21 %	48.43 %	0.0004	99.61 %
113	11881.82	0.0007	55.62 %	23.05 %	56.38 %	18.79 %	37.58 %	0.0003	99.68 %
114	11985.96	0.0005	77.21 %	23.88 %	77.80 %	25.93 %	51.87 %	0.0002	99.73 %
115	12090.1	0.0005	137.56 %	24.72 %	137.91 %	45.97 %	91.94 %	0.0005	99.77 %
116	12194.24	0.0004	155.18 %	25.55 %	155.52 %	51.84 %	103.68 %	0.0004	99.82 %
117	12298.38	0.0003	138.02 %	26.38 %	138.43 %	46.14 %	92.28 %	0.0003	99.85 %
118	12402.52	0.0002	78.21 %	27.22 %	78.96 %	26.32 %	52.64 %	0.0001	99.87 %
119	12506.66	0.0002	210.84 %	28.05 %	211.14 %	70.38 %	140.76 %	0.0003	99.89 %

Footnotes

1. Sum of square of all types of uncertainties to β_i^0 including type A and type B uncertainties.

2. Following NIST policy the standard uncertainty of the mean is multiplied by 2 to get an estimate of uncertainty limits.[24]

Table 4

The fraction, β_i^0, of each oligomer in the MMD and the uncertainty in β_i^0 for SRM 2881

repeat number, i	Oligomer mass, m_i	MMD fraction β_i^0	Uncertainty in β_i^0 using coverage factor of k=2	Cumulative Distribution
62	6570.68	0.0018	0.0005	1.0 %
63	6674.82	0.0020	0.0007	1.2 %
64	6778.96	0.0025	0.0007	1.5 %
65	6883.1	0.0031	0.0009	1.8 %
66	6987.24	0.0041	0.0010	2.2 %
67	7091.38	0.0048	0.0013	2.7 %
68	7195.52	0.0063	0.0014	3.3 %
69	7299.66	0.0078	0.0014	4.1 %
70	7403.8	0.0090	0.0016	5.0 %
71	7507.94	0.0116	0.0018	6.1 %
72	7612.08	0.0135	0.0019	7.5 %
73	7716.22	0.0163	0.0019	9.1 %
74	7820.36	0.0191	0.0018	11.0 %
75	7924.5	0.0223	0.0019	13.3 %
76	8028.64	0.0262	0.0017	15.9 %
77	8132.78	0.0296	0.0017	18.8 %
78	8236.92	0.0322	0.0013	22.1 %
79	8341.06	0.0353	0.0011	25.6 %
80	8445.2	0.0385	0.0014	29.4 %
81	8549.34	0.0404	0.0012	33.5 %
82	8653.48	0.0423	0.0012	37.7 %
83	8757.62	0.0440	0.0013	42.1 %
84	8861.76	0.0446	0.0017	46.6 %
85	8965.9	0.0464	0.0018	51.2 %
86	9070.04	0.0446	0.0019	55.7 %
87	9174.18	0.0440	0.0019	60.1 %
88	9278.32	0.0424	0.0019	64.3 %
89	9382.46	0.0404	0.0018	68.3 %
90	9486.6	0.0380	0.0017	72.1 %
91	9590.74	0.0359	0.0018	75.7 %
92	9694.88	0.0325	0.0017	79.0 %
93	9799.02	0.0297	0.0016	82.0 %
94	9903.16	0.0268	0.0015	84.6 %
95	10007.3	0.0242	0.0015	87.1 %
96	10111.44	0.0207	0.0013	89.1 %
97	10215.58	0.0182	0.0010	91.0 %
98	10319.72	0.0155	0.0014	92.5 %

99	10423.86	0.0139	0.0008	93.9 %
100	10528	0.0113	0.0008	95.0 %
101	10632.14	0.0096	0.0005	96.0 %
102	10736.28	0.0077	0.0005	96.8 %
103	10840.42	0.0063	0.0004	97.4 %
104	10944.56	0.0051	0.0005	97.9 %
105	11048.7	0.0042	0.0002	98.3 %
106	11152.84	0.0031	0.0004	98.6 %
107	11256.98	0.0026	0.0002	98.9 %
108	11361.12	0.0019	0.0012	99.1 %
109	11465.26	0.0020	0.0004	99.3 %
110	11569.4	0.0013	0.0004	99.4 %
111	11673.54	0.0011	0.0004	99.5 %
112	11777.68	0.0009	0.0003	99.6 %
113	11881.82	0.0007	0.0002	99.7 %
114	11985.96	0.0005	0.0005	99.7 %
115	12090.1	0.0005	0.0004	99.8 %
116	12194.24	0.0004	0.0003	99.8 %
117	12298.38	0.0003	0.0001	99.8 %
118	12402.52	0.0002	0.0003	99.9 %
119	12506.66	0.0002	0.0005	99.9 %

Appendix A

National Institute of Standards & Technology

Certificate

Standard Reference Material® 2881

Polystyrene

This Standard Reference Material (SRM) is intended primarily for use in the calibration and the performance evaluation of instruments used to determine the molar mass and molar mass distribution of synthetic polymers. These methods include size exclusion chromatography (SEC) [1] and mass spectrometry (MS) [2]. The fractional molar mass of each oligomer from 1 % to 99 % of the cumulative mass distribution were certified and are given in Table 1

A unit of SRM 2881 consists of approximately 0.2 g of polystyrene powder initiated with an n-octyl group and terminated with a proton.

Certified Uncertainties: The certified measurement uncertainty is expressed as a combined expanded uncertainty with a coverage factor $k = 2$, calculated in accordance with NIST procedure [3]. Type A and Type B contributions to the expanded uncertainty of the measured molar mass fractions include the uncertainties in instrument and sample preparation methods.

Expiration of Certification: The certification of SRM 2881 is valid, within the measurement uncertainties specified, until 1 January 2019, provided that the SRM is handled in accordance with the storage instructions given below (see "Instructions for Use"). This certification is nullified if the SRM is modified, contaminated, or stored improperly.

Maintenance of SRM Certification: NIST will monitor this SRM over the period of its certification. If substantive technical changes occur that affect the certification before expiration of this certificate, NIST will notify the purchaser. Return of the attached registration card will facilitate notification.

Supplemental Information: M_n was determined by nuclear magnetic resonance (NMR) and found to be $9.1 \times 10^{+3}$ g/mol with an estimated uncertainty of $\pm 0.50 \times 10^{+3}$ g/mol. NMR and matrix-assisted laser desorption/ionization time-of-flight mass spectrometry (MALDI-TOF-MS) were used to analyze end groups on the polymer. Only the set of end groups (as specified above) were found.

NIST Certification Method: The certified value for the oligomeric fractions in the molecular mass distribution was measured using matrix-assisted laser desorption/ionization time-of-flight mass spectrometry (MALDI-TOF MS) [4,5].

INSTRUCTIONS FOR USE

Storage: The SRM should be stored in the original bottle with the lid tightly closed under normal laboratory conditions.

Homogeneity and Characterization: The homogeneity of SRM 2881 was tested by SEC analysis of solutions in tetrahydrofuran at 35 °C. The further characterization of this polymer is described in reference [4, 5].

William E. Wallace of the NIST Polymers Division provided technical coordination leading to certification of this SRM.

Charles M. Guttman, William E. Wallace, Kathleen M. Flynn, and David L. VanderHart of the Polymers Division provided technical measurement and data interpretation. Anthony J. Kearsley of the Mathematical and Computational Sciences Division provided methodology for instrument optimization and for unbiased peak integration methods for the data.

Statistical consultation was provided by John Lu of the Statistical Engineering Division.

Support aspects involved in the preparation and issuance of this SRM were coordinated through the Measurement Services Division.

Eric K. Lin, Chief
Polymers Division

Gaithersburg, MD 20899
Certificate Issue Date: 28 October 2008

Robert L. Watters, Jr., Chief
Standard Reference Materials Program

Table I

The number fraction β_i for each oligomer in the molecular mass distribution with associated uncertainties for SRM 2881

Repeat Unit Number	Oligomer Mass	Number Fraction of MMD	Uncertainty in Number Fraction	Cumulative Percentage of MMD
i	g/mole	β_i	k=2	%
63	6675.80	0.0020	0.0003	1.0%
64	6779.95	0.0025	0.0005	1.2%
65	6884.10	0.0033	0.0006	1.6%
66	6988.26	0.0041	0.0007	2.0%
67	7092.41	0.0053	0.0008	2.5%
68	7196.56	0.0067	0.0010	3.2%
69	7300.71	0.0084	0.0011	4.0%
70	7404.86	0.0105	0.0013	5.1%
71	7509.01	0.0129	0.0014	6.4%
72	7613.17	0.0157	0.0016	7.9%
73	7717.32	0.0186	0.0017	9.8%
74	7821.47	0.0217	0.0017	12.0%
75	7925.62	0.0254	0.0017	14.5%
76	8029.77	0.0287	0.0017	17.4%
77	8133.93	0.0321	0.0017	20.6%
78	8238.08	0.0353	0.0016	24.1%
79	8342.23	0.0383	0.0015	28.0%
80	8446.38	0.0410	0.0011	32.1%
81	8550.53	0.0431	0.0010	36.4%
82	8654.69	0.0449	0.0007	40.8%
83	8758.84	0.0456	0.0007	45.4%
84	8862.99	0.0459	0.0006	50.0%
85	8967.14	0.0457	0.0008	54.6%
86	9071.29	0.0450	0.0010	59.1%
87	9175.44	0.0437	0.0012	63.4%
88	9279.60	0.0416	0.0014	67.6%
89	9383.75	0.0397	0.0016	71.6%
90	9487.90	0.0368	0.0017	75.2%
91	9592.05	0.0339	0.0017	78.6%
92	9696.20	0.0310	0.0017	81.7%
93	9800.36	0.0279	0.0016	84.5%
94	9904.51	0.0249	0.0016	87.0%
95	10008.66	0.0219	0.0015	89.2%
96	10112.81	0.0191	0.0014	91.1%
97	10216.96	0.0164	0.0013	92.7%
98	10321.12	0.0140	0.0011	94.1%
99	10425.27	0.0118	0.0010	95.3%
100	10529.42	0.0098	0.0009	96.3%
101	10633.57	0.0081	0.0007	97.1%
102	10737.72	0.0066	0.0006	97.8%
103	10841.87	0.0053	0.0006	98.3%
104	10946.03	0.0042	0.0005	98.7%
105	11050.18	0.0034	0.0005	99.1%

REFERENCES

[1] See ASTM D5296 05 "Standard Test Method for Molecular Weight Averages and Molecular Weight Distribution of Polystyrene by High Performance Size-Exclusion Chromatography" and International Organization for Standardization (ISO) 16014-1:2003.

[2] See ASTM D7034 05 "Standard Test Method for Molecular Mass Averages and Molecular Mass Distribution of Atactic Polystyrene by Matrix Assisted Laser Desorption/Ionization (MALDI) Time of Flight (TOF) Mass Spectrometry (MS)", Deutsches Institut für Normung DIN 55674, and International Organization for Standardization (ISO) CD 10927.

[3] B.N. Taylor, C.E. Kuyatt, "Guidelines for Evaluating and Expressing the Uncertainty of NIST Measurement Results," NIST Technical Note 1297, U.S. Government Printing Office. Washington D.C., 1994.

[4] C.M. Guttman, K.M. Flynn, W.E. Wallace, A.J. Kearsley, Report on the Certification of an Absolute Molecular Mass Distribution Polymer Standard: Standard Reference Material 2881, NIST Internal Report 7512, August 2008.

[5] C.M. Guttman, K.M. Flynn, W.E. Wallace, A.J. Kearsley, Quantitative Mass Spectrometry and Polydisperse Materials: Creation of an Absolute Molecular Mass Distribution Polymer Standard, *Macromolecules* (submitted).

www.ingramcontent.com/pod-product-compliance
Lightning Source LLC
Chambersburg PA
CBHW081859170526
45167CB00007B/3078